Digital Thirst and Climate Change: What does boom in resource-hungry data centres mean for climate change and how to Navigating the Future of Energy in the Age of Data Centers.

By

Roger Munoz

Copyright©2024 Roger Munoz

All rights reserved. No part of this book may be reproduced or modified in any form, including photocopying, recording, or by any information storage and retrieval system, without permission in writing from the publisher.

DEDICATION

I dedicate this book to all my wonderful and amazing readers.

Thank you so much!

ABOUT THE AUTHOR

Roger Munoz is an esteemed American writer and researcher, renowned for his insightful explorations into the intersections of climate change, technology, and business. With a career spanning over two decades, Munoz has established himself as a pivotal voice in contemporary discourse, offering both critique and solutions to some of the most pressing issues of our time.

Hailing from a modest background, Munoz's journey into the world of writing and research was driven by an insatiable curiosity about the world and a fervent desire to make a difference. After earning a degree in Environmental Science from a prestigious university, he ventured further, acquiring a Master's in Business Administration, which equipped him with a unique lens to view the complexities of modern societal challenges.

SOME CLIMATE CHANGE NUGGETS

Oceanic Heatwaves: Just like on land, the ocean experiences heatwaves, too. These marine heatwaves disrupt marine ecosystems, causing species to migrate to cooler waters and affecting the food chain.

Invisible Methane: While CO_2 gets most of the attention, methane is over 25 times more potent in trapping heat in the atmosphere over a 100-year period, making its reduction critical to slowing global warming.

Greenland's Vanishing Ice: Greenland's ice sheet is melting six times faster now than in the 1980s. This dramatic change contributes to rising sea levels, threatening coastal communities worldwide.

Renewable Revolution: Solar and wind energy prices have plummeted over the past decade, making them cheaper than fossil fuels in many parts of the world and offering a beacon of hope for reducing carbon emissions.

Climate Refugees: Climate change is creating millions of climate refugees—people forced to leave their homes due to rising seas, extreme weather, and disrupted agriculture.

Forest as Lungs: Forests are often described as the Earth's lungs, absorbing CO_2 and releasing oxygen. The Amazon rainforest alone produces 20% of the world's oxygen.

Coral Bleaching Crisis: Warmer oceans lead to coral bleaching, where corals lose the algae that feed them and give them color. This disrupts marine habitats and the livelihoods of people who depend on them.

Permafrost Peril: The thawing of permafrost—ground that remains completely frozen for two or more years—

in the Arctic releases ancient methane and CO2, further amplifying global warming.

Eco-Anxiety: As the impacts of climate change become more visible, eco-anxiety—worry about the environmental crisis—is rising, affecting mental health across all ages.

Faster Than Predicted: Climate change is happening at a pace faster than many scientists predicted. The last decade was the warmest on record, emphasizing the urgent need for action.

Urban Heat Islands: Cities often experience higher temperatures than their rural surroundings, a phenomenon known as the urban heat island effect. This increases energy consumption for cooling and exacerbates health problems.

Biodiversity at Risk: Climate change is a key driver of biodiversity loss. As habitats change or disappear, many species struggle to survive, reducing the resilience of natural ecosystems and their ability to support human life.

TABLE OF CONTENT

DEDICATION 2

ABOUT THE AUTHOR 3

SOME CLIMATE CHANGE NUGGETS...4

TABLE OF CONTENT 7

INTRODUCTION 11

THE RISE IN SILICON CONSUMPTION 11
ANALYZING THE DATA BOOM FROM AN INTERNATIONAL PERSPECTIVE 12
THE HIDDEN COST OF CLICKS: THE ENERGY BENEATH THE SCREEN......................13

CHAPTER ONE..17

FOUNDATIONS OF THE DIGITAL AGE 17
THE DATA CENTER'S HISTORY..18
SILICON VALLEYS: THE GEOGRAPHIC DISTRIBUTION OF THE DATA 19
THE TRANSITION FROM MAINFRAMES TO CLOUD 20

CHAPTER TWO...23

ENERGIZING THE INVISIBLE: A SUMMARY OF ENERGY USE IN DATA CENTERS.........24

Internet Power Mix: Non-Renewable vs. Renewable Energy 25
Vacuum Filling 27
The Way Forward 28

CHAPTER THREE 29

The Environmental Paradox 29
Carbon Footprints in the Cloud: The Environmental Impact of Online Storage 30
Fresh Methods for Reducing Data Center Heat: Cooling the Giants 31
Resolving the Conundrum 33
The Way Forward 33

CHAPTER FOUR 35

....35
The Economic Engine 35
The Cost of Connectivity: The Economic Impact of Data Centers 36
Payoffs and Invested Capital: Finance for the Data Future 37
Connecting the Dots 39

CHAPTER FIVE 41

Policy and Power 41
Supervision and Responsibility: Monitoring the Advancement of Digitization 42
International Views on Data Center Energy: Worldwide Discussions 44
The Path Ahead: Striking a Balance Between Innovation and Sustainability 45

CHAPTER SIX 47

THE FUTURE OF DATA CENTERS..47
MONITORING THE DIGITIZATION PROCESS WITH ACCOUNTABILITY AND CONTROL ..48
INTERNATIONAL PERSPECTIVES ON DATA CENTER ENERGY IN GLOBAL DISCUSSIONS49
THE WAY FORWARD: BALANCING INNOVATION AND SUSTAINABILITY....................51

CHAPTER SEVEN 53

SOCIETAL IMPLICATIONS 53
THE DATA DIVIDE'S DIGITAL ECONOMY AND ACCESSIBILITY 54
POWER, DATA, AND PRIVACY: HANDLING THE NEW NORMAL...............................55
BUILDING BRIDGES AND SAFEGUARDING SANCTUARIES..56

CHAPTER EIGHT 59

TOWARDS A SUSTAINABLE DIGITAL ECOSYSTEM 59
EMBRACING RENEWABLE ENERGY: LEARNINGS FROM CASE STUDIES OF GREEN
CHANGE 60
AI AND MACHINE LEARNING AS A GREEN CATALYST 61
CHALLENGES AND PROSPECTIVE PATHS 62
THE COMBINED JOURNEY 63
TO SUM UP, A FUTURE OUTLOOK..63

CHAPTER NINE 65

IMPACT OF DATA CENTER BOOM ON CLIMATE CHANGE 65
EXAMINING THE ENVIRONMENTAL EFFECTS: UNDERSTANDING THE RELATIONSHIP
BETWEEN DATA CENTERS AND CLIMATE CHANGE 66
GETTING THROUGH THE STORM: METHODS FOR MITIGATING DATA CENTER-
RELATED CLIMATE IMPACTS 67

CHAPTER TEN **71**

STRATEGIES FOR A SAFER GLOBAL ENVIRONMENT..**71**
UNDERSTANDING THE TRADE-OFFS: GROWTH VS. SUSTAINABILITY......................72
EMBRACING SUSTAINABLE PRACTICES: AN ALL-ENCOMPASSING APPROACH..........73
REAPING THE BENEFITS OF RENEWABLE ENERGY..74
ENCOURAGING INNOVATION AND COLLABORATION 75
CONCLUDING DISCUSSION: BUILDING A SUSTAINABLE FUTURE............................76

INTRODUCTION

The emergence of the digital age has led to an unprecedented need for data, fundamentally changing our way of life, work, and social connections. A humble but crucial component, silicon sits at the heart of this transformation. This chapter examines the rise in global data consumption as well as the often-overlooked energy costs of our digital activities.

The Rise in Silicon Consumption

Today, silicon—the second most prevalent element in Earth's crust—serves as the foundation for modern technology. It is the primary material for integrated circuits, which power everything from servers in data centers to cellphones, due to its remarkable semiconducting characteristics. Our reliance on digital products and services has raised our need for silicon.

Every message sent, every video watched, and every search term entered appears to contribute just a little bit more to the rapidly increasing demand for digital content.

There have been consequences to this explosion. The complicated chip fabrication and silicon purification processes involved in silicon chip manufacture require precision and considerable energy usage. More water and energy are used throughout each step than most people would think, contributing to a greater environmental impact than most people realize. The complicated, resource-intensive process that begins in mines and ends with microchips produces our everyday devices.

Analyzing the Data Boom from an International Perspective

Transcending geographical borders and cultural disparities, the data explosion is an essentially worldwide phenomenon. It is propelled by the expanding number of internet users as well as the expanding Internet of Things (IoT), a network of intelligent devices that collects and sends data to make the world more connected. Consider how a smart thermostat in a Tokyo apartment may discover its owner's preferences or how a Kenyan farmer might use a smartphone app to monitor crop prices in his rural area. Due to the rising demand for data globally, it ultimately results in an increase in the requirement for servers, storage, and energy.

This growth is remarkable. Data centers, the powerhouses of the internet, now consume 1% of the world's electricity, a number that is increasing yearly. These hubs need to be continuously running and ready to service our digital demands, whether it's a teenager in Brazil streaming a movie or a doctor in Germany getting access to a patient's medical records.

The Hidden Cost of Clicks: The Energy Beneath the Screen

Beyond the confines of our devices, every click, swipe, or voice command initiates a cascade of actions that reach deep into the enormous data centers that contain and manage our digital needs. The cooling systems in these facilities consume a significant amount of energy in order to prevent overheating. These activities typically rely on fossil fuels as their energy source, which raises greenhouse gas emissions and accelerates climate change.

Watch a movie on streaming in high definition. Though it looks like a simple, low-impact activity, it's everything but. The data for the video is stored in a data center that could be thousands of kilometers away. Servers use a lot of electricity to deliver this data over great distances throughout the streaming process. This kind of action requires a significant amount of energy when millions of users do it at once.

But rather than being a criticism of their application, this is a call to acknowledge and deal with the environmental impact of digital technology. The solution is to make digital services more energy-efficient and to derive their electricity from renewable sources rather than abandoning them. Improvements in server performance, advanced cooling techniques, and the transition to greener energy sources are examples of technological progress that is already making an impact.

Understanding the link between our digital activities and their energy use is the first step towards a more sustainable digital future. It's about understanding that our seemingly trivial activities are only one part of the larger picture of energy usage and environmental effect. Consumer preferences and expectations for eco-friendly practices can have an impact on the sector.

There are responsibilities and risks associated with the surge in silicon consumption brought on by our insatiable appetite for data. By considering data consumption from a global perspective and the hidden costs of our digital existence, we may begin to appreciate the complexity of these issues and the importance of finding solutions that strike a balance between the health of our planet and our technological advancements. Gaining a deeper understanding of this will enable us to navigate the digital landscape and ensure that environmental sustainability does not suffer as a result of our growth.

CHAPTER ONE

Foundations of the Digital Age

The digital era has had a revolutionary history marked by rapid advancements and significant shifts in how we utilize technology. The fundamental focus of this story is the evolution of data processing and storage from enormous mainframes of the past to sleek, ubiquitous cloud services of the present. This journey not only altered the technological landscape but also the physical and environmental limits of data storage.

The Data Center's History

Once upon a time, computers were massive, room-sized devices that processed information at a speed that is now considered slow while humming and blinking. These were the mainframes from the mid-1900s, the forerunners of today's data centers. Mainframes, which were centralized hubs containing all of the computing power, were the earliest type of data centers. Businesses and governmental institutions processed data using these massive devices, which was expensive and time-consuming.

Nevertheless, inventions are born out of need. The limitations of mainframes—their size, cost, and inefficiency—created a need for more readily available, scalable alternatives. Greetings from the era of personal computers and servers. The decentralization from a few giant mainframes to multiple smaller servers was a significant first step toward the creation of modern data centers. These servers needed a place to reside despite their lower size, which led to the creation of the first data centers—areas designated especially for processing and storing data.

Silicon Valleys: The Geographic Distribution of the Data

There was an increasing need for data processing and storage as digital technology advanced. This necessity led to the emergence of densely populated data center regions. These areas are known as "Silicon Valleys" in jest because to the high concentration of silicon-powered devices found there. Unlike the distinctive Silicon Valley in California, which is well-known for its digital innovation and businesses, these "Silicon Valleys" of data are scattered over the globe, from the desert parts of Arizona to the freezing regions of Northern Europe.

These places were chosen with practical factors in mind. Lower temperatures, for instance, reduce the need for artificial cooling in data centers, saving a significant amount of energy. In a similar vein, places with easy access to renewable energy sources, such the hydroelectric power in the Pacific Northwest, developed became hubs for data center building. The geographical distribution of these data centers highlights the shift from physical, concentrated sources of computing power to a dispersed but connected network, highlighting the global nature of the digital era.

The Transition from Mainframes to Cloud

The move from physical servers to cloud computing is the most recent advancement in data processing and storage. Cloud computing allows data to be stored and retrieved over the internet, releasing users from the constraints of physical hardware. Because of this modification, data is now more readily available, scalable, and reasonably priced. Instead of buying and maintaining actual servers, companies and individuals can now rent processing and storage capacity from cloud service providers, paying only for what they use.

Imagine a small startup with limited capital. In the past, setting up a server for a fresh application could have proved disastrous. By putting their application in the cloud, they may now instantly access the same amount of processing power that belonged to giants in the industry like Google and Amazon.

 Since technology is now more widely available, there is more potential than ever for innovation and development.

The cloud signifies a shift in our perception of and interactions with data that goes beyond a purely technological one.

Data is now everywhere and always available, whether we're using a laptop halfway across the world to examine work files or viewing a movie on our smartphone. One of the defining characteristics of the digital age is the pervasiveness of data due to the cloud.

CHAPTER TWO

The Energy Backbone of the Digital World

The faultless digital experiences we cherish, like making purchases online, video chatting with friends across the globe, or binge-watching our favorite program, are powered by a massive hidden energy core. This digital world's engine is this backbone, primarily made up of data centers. Understanding the energy usage of these data centers and the mix of non-renewable and renewable sources that power the internet is essential if we are to properly appreciate the sustainability opportunities and challenges we confront in the digital age.

Energizing the Invisible: A Summary of Energy Use in Data Centers

Imagine for a moment, day or night, a perpetually lit metropolis. This city is like the biggest data center network in the world, the beating heart of the internet, where every search query, streaming video, and cloud-stored photo requires an electrical pulse. Data centers are big structures with a lot of servers that distribute, process, and store data. Despite their unseen powering of our digital life, they consume an enormous amount of energy.

Data centers need an incredible amount of electricity. To give you a perspective of scale, one data center can use as much electricity as tens of thousands of homes. Both the servers themselves and the cooling equipment needed to dissipate the heat they generate necessitate this requirement. Because high temperatures can damage equipment and reduce its efficiency, which can lead to data loss or interrupted services, cooling is crucial.

Imagine your laptop overheating and the performance drop that results until it cools down to put it in a relatable context. You may understand why data centers require so much energy for cooling when you take into account the thousands of servers operating in a compact space.

Internet Power Mix: Non-Renewable vs. Renewable Energy

The source of the electricity used by data centers determines the environmental impact of our digital activity. This energy has traditionally been produced by non-renewable resources including coal, natural gas, and oil. These sources are not only scarce, but they also contribute significantly to the greenhouse gas emissions that drive global warming.

Still, a radical shift is taking place. Renewable energy sources like solar, wind, and hydroelectric electricity are increasingly being used to power data centers. These clean, abundant, and sustainable sources offer a path forward for a digital future that is more ecologically benign. In addition to combating climate change, switching to renewable energy sources will assist maintain the long-term viability of the digital infrastructure supporting contemporary society.

Consider a data center that uses photovoltaic farms to generate its electricity. The data center will be able to run entirely on saved solar energy thanks to recent advancements in battery storage technology. This case study highlights the value of innovation and green technology investment in addition to demonstrating that data centers may be fueled by renewable energy sources.

Vacuum Filling

Making the switch from a primarily non-renewable energy-based internet to one that is powered only by renewable energy is a challenging task. Governments, corporations, and people all need to collaborate, make significant investments, and exercise creativity. Every single stakeholder matters. For instance, people can support firms and policies that prioritize sustainability, governments can provide incentives for green energy projects, and IT companies can commit to running their entire organization entirely on renewable energy.

One positive real-world example is a digital company that made global investments in renewable energy projects to achieve carbon neutrality for its data centers. In addition to offsetting its energy use, this contributes to the global expansion of the infrastructure for renewable energy sources.

The Way Forward

The digital world's energy infrastructure will be scrutinized more and more. The transition to renewable energy is not only essential for the environment but also has the ability to progress technology and create jobs. Making the switch to a greener internet is a big task, but it also offers an opportunity to build a more reliable and long-lasting digital infrastructure.

It is still possible to have a sustainable digital future in spite of the challenges. Comprehending the energy dynamics of data centers and advocating for the transition to sustainable energy sources will contribute to ensuring that the digital world of the future is not just bright and growing, but also less detrimental to the environment. In the unfolding story about the energy infrastructure of the digital world, everyone has a part to play.

Roger Munoz

CHAPTER THREE

The Environmental Paradox

In the realm of digital technology, we face an astounding paradox: the very technologies designed to make our lives easier and reduce physical waste simultaneously have a significant detrimental impact on the environment, particularly in terms of energy use and the resulting carbon footprints. The hidden environmental costs of our online behavior are examined in this chapter, with a focus on online storage and the inventive methods being devised to cool down the data centers—the giants of our digital age.

Carbon Footprints in the Cloud: The Environmental Impact of Online Storage

Cloud-stored data has a transient, airy feel about it, as though it were floating in midair. As a matter of fact, "the cloud" is actually composed of enormous data centers that consume copious amounts of electricity, a large portion of which has historically come from fossil fuels. Every email received, photo uploaded, and document stored increases the data centers' carbon footprints.

The environmental impact of online storage becomes more evident when you consider the massive volume of data being stored. Millions of users, for example, stream videos every day, which implies that data servers storing these movies must always have power and data must always be retrieved. This process releases a significant amount of carbon dioxide, which is the primary cause of climate change. In order to put things into perspective, consider that if the internet were a country, its electricity demand would rank sixth in the world. This illustrates the massive energy consumption and associated carbon emissions that come with online activity.

Fresh Methods for Reducing Data Center Heat: Cooling the Giants

The heat produced by the servers is one of the main issues in a data center. It may result in device damage, reduced efficacy, and overheating if left unchecked. Traditional cooling methods include air conditioning equipment, which increases energy use and carbon emissions even further. Nevertheless, the industry is developing, searching for strategies to reduce temperatures with less detrimental effects on the environment.

Among the most intriguing developments in this subject is the use of natural cooling technology. These days, a few data centers are located in cooler climates where the outside air can naturally cool the servers, significantly reducing the need for air conditioning. For instance, a data center in Northern Europe uses the chilly Arctic air to maintain optimal temperatures, saving a substantial amount of electricity.

Utilizing submerged data centers is an additional cutting-edge tactic. Companies can submerge server farms under the ocean to take use of the natural cooling properties of seawater. This reduces the need for artificial cooling, and opens up new possibilities for sustainable energy use when combined with wave and tidal energy generation.

Furthermore, the development of heat recovery technologies is a significant step toward sustainability. These solutions, which repurpose server heat for building heating, further reduce the environmental effect of data centers. Imagine an office building where the company's data servers, which manage the flow of information, act as heaters in the winter. This is the perfect example of using a challenge to your advantage.

Resolving the Conundrum

Digital technology's environmental conflict has remedies. At the moment, renewable energy sources—solar and wind—are the mainstay for powering data centers. This shift is necessary to reduce the carbon footprint of online storage and the larger internet infrastructure.

Education and transparency are also crucial. As more individuals become aware of how their online behavior affects the environment, there is an increasing demand for sustainable behaviors. The desire from customers drives companies to keep innovating in areas like energy efficiency and cooling techniques, as well as to invest in green technologies.

The Way Forward

The continuous journey towards a sustainable digital infrastructure is replete with both opportunities and challenges. Developments in cooling technologies and the increasing use of renewable energy sources are encouraging signs of progress. However, considering the size of the issue, collaboration between IT firms, governmental bodies, and end users is imperative.

Comprehending the ecological conundrum posed by our digital behaviors enables both individuals and society to make more informed choices. The environmental impact of the digital world is being lessened by every step made to support green technology regulations or enterprises that prioritize sustainability.

The fate of our planet will depend on our ability to strike a balance between the benefits of digital technology and the need for environmental care. As long as we continue to innovate and adapt, there is optimism that we can resolve this paradox and ensure that our digital advances serve society both online and offline.

CHAPTER FOUR

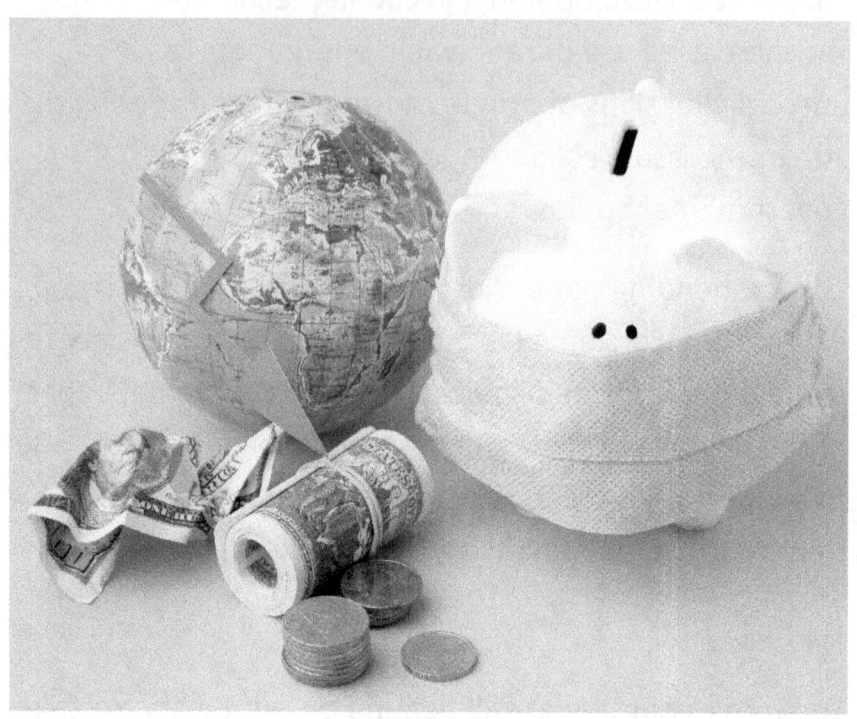

The Economic Engine

As we go deeper into the digital era, the data centers that power our online activities start to emerge as technological marvels as well as significant players in the global economy. This chapter looks at data centers from two angles: the significant connection costs and the developing possibilities for funding future data through incentives and investments.

The Cost of Connectivity: The Economic Impact of Data Centers

To understand the financial aspects of data centers, let's start with a basic need of the digital age: connectivity. Every tweet, every swipe on a dating app, and every online transaction go swiftly through the global network of data centers. This continuous connectedness puts the world at our fingertips, but at a high cost.

creating a small power plant is akin to creating and operating a data center. One has to factor in the cost of sophisticated equipment like computers, cooling systems, and backup power supply in addition to the land and building prices. And that's before you even factor in the exorbitant cost of electricity. The yearly budget of a small municipality, for example, might be similar to what it costs to run a mid-sized data center.

However, the impact of these costs on the economy extends beyond consumer expenditure. It's also a story of technological advancement, expanding infrastructure, and job development. Since the construction, operation, and maintenance of data centers require skilled people, a significant number of high-tech jobs are created by them. Furthermore, the existence of data centers might attract additional industries and businesses, creating a tech cluster that fosters local economic growth. Just as flowers surround a well-fed plant, so too do other establishments—restaurants, shops, and services—bloom all around a rising data center. We call this the "ripple effect."

Payoffs and Invested Capital: Finance for the Data Future

The financing of the future of data requires a thorough understanding of the variety of investments and incentives that allow data centers to expand and innovate. The significance of building a robust digital infrastructure is acknowledged by both the public and private sectors, as evidenced by their financial investments.

Incentives play a crucial role in this ecology. Tax breaks, grants, and subsidies are common tools used by governments to lure data center building to certain regions. These incentives are not just arrogant presents given to digital behemoths; they are calculated wagers on the future economic success of an area. By offering these benefits, a place can draw in data centers more easily. As we've shown, these benefits can work as catalysts for broader economic growth.

The story is equally compelling from the standpoint of investing. Data centers are seen by investors as excellent investments with steady, long-term return potential. Its allure is aided by the fact that data processing and storage capacity are always needed. Pension funds, venture capital firms, and private equity firms provide finance for data center building projects. The logic behind this is simple: as our lives get more digital, so does the infrastructure that supports and facilitates digital living.

An excellent illustration of this is a tiny town that almost overnight became a tech hub due to strategically placed investments in data center infrastructure. The town's economy was initially focused on traditional sectors, but once a sizable data center moved in, startups and digital enterprises started to see it as a viable place to call home. This shift is a good illustration of how to finance the data-driven future because it was made possible by a combination of government incentives and private investment.

Connecting the Dots

Despite their high cost, data centers are a complex economic engine that offer vast economic opportunities. Understanding this engine requires striking the correct balance between the expenses of creating and maintaining this digital infrastructure and the benefits it brings to the community's economy in terms of growth, innovation, and jobs.

As time goes on, data centers will become more and more significant to the economy. Thanks to developments in edge computing, artificial intelligence, and renewable energy, the data center industry is poised to increase in effectiveness, strength, and significance as a part of our daily lives. Our future, however, will depend on our ability to continue financing this expansion through a combination of government grants, private investment, and innovative financing techniques.

Data centers are situated at the intersection of technology and the economy, acting as both generators and recipients of economic expansion. Their story, which illustrates the transformative potential of digital infrastructure in transforming future economic settings, is one of immense possibility and great expenditure. Making sure that this economic engine drives prosperity and inclusive progress for all, not just a select few, will be the challenge that lies ahead.

CHAPTER FIVE

Policy and Power

In the intricate dance between technology and society, the data center industry offers a unique choreography of accountability and control. This chapter examines the ways in which governments, businesses, and communities are managing the expansion of digital infrastructure. It draws attention to significant regulatory concerns as well as the various international approaches to managing data center energy consumption.

Supervision and Responsibility: Monitoring the Advancement of Digitization

The digital age is defined by a paradox: the very technologies that connect us, foster innovation, and advance economies yet need massive amounts of energy and have a significant detrimental effect on the environment. In this instance, regulation is quite important. It all boils down to finding a balance between environmental preservation and technological advancement.

Assume you are creating a city plan. Roads are necessary in addition to parks. In a similar vein, rules are required as our digital world develops to ensure that environmental costs are minimized. Regulations have the power to do everything from establish energy efficiency standards to place restrictions on the use of renewable energy sources.

For example, a country may pass laws mandating that data centers reveal their energy consumption and carbon emissions. This is similar to encouraging drivers to check the emissions from their vehicle; it's a positive step in identifying the problem's extent and creating solutions. Some communities even go so far as to offer tax discounts to data centers that satisfy certain environmental requirements—benefits that are comparable to those offered to owners of electric vehicles.

However, enforcing regulations is a difficult task. A rigorous balancing act is required to achieve significant change through strict regulations without impeding innovation or driving the industry towards less regulated sectors. It serves the same purpose as speed limits, protecting people without making them late for work.

International Views on Data Center Energy: Worldwide Discussions

A variety of strategies for regulating data center energy use are available on the global stage, each of which takes into account the unique needs and preferences of various nations. Rich in renewable resources, certain countries support data centers fueled by green energy, while energy-poor countries prioritize efficiency and innovation.

Think about Iceland, where geothermal and hydroelectric electricity allow data centers to operate almost totally on sustainable energy. This is like having all of your electricity coming from solar panels on your roof and a wind turbine in your garden. Despite its limited space and lack of renewable energy sources, Singapore, on the other hand, uses leading edge technologies such as undersea data centers that use saltwater for cooling, minimizing the need for energy-intensive air conditioning.

Collaborations and international agreements are also included. Collaboration is becoming more and more necessary to manage and control data center energy use, much like nations are banding together to combat climate change. Think of it as a neighborhood group that decides to combine its resources and tools to improve their homes' energy efficiency.

Addressing the transboundary nature of data flows and the internet, as well as exchanging best practices, requires an international dialogue. After all, national boundaries do not apply to data. Users who watch videos hosted on servers in another country while in another country offer daily proof of this interconnectedness. Therefore, international collaboration is crucial in the digital domain just as it is in addressing global concerns like climate change.

The Path Ahead: Striking a Balance Between Innovation and Sustainability
Regulating and controlling data center energy is an ongoing effort, much like urban planning for a city that is always growing. It requires foresight, flexibility, and a willingness to adjust to new developments in technology.

Future developments are likely to bring stricter regulations, more inventive energy solutions, and increased international cooperation. The parts are present, but the issue is challenging. Data centers can be constructed sustainably and made more efficient with the use of technology and regulatory tools. The challenge lies in combining these elements in a way that balances the needs of our data-hungry society with the health of our planet.

Basically, as we continue to create the digital infrastructure of the future, we must take into account both the power and the responsibilities that accompany it. Data centers are about more than just bandwidth and bytes; they are about using these resources to create a society that functions both online and offline. It's about ensuring that, while our digital development soars, the planet on which we live is not endangered.

Roger Munoz

CHAPTER SIX

The Future of Data Centers

The data center sector offers a unique choreography of accountability and control in the complex dance between technology and society. This chapter examines the development of digital infrastructure and how it affects governments, businesses, and communities. The essay highlights important legal issues as well as different international approaches to controlling the energy use of data centers.

Monitoring the Digitization Process with Accountability and Control

The digital age is characterized by a paradox: the very technologies that promote economic expansion, encourage innovation, and link people also use enormous quantities of energy and negatively impact the environment. In this case, regulation is crucial. It's critical to strike a balance between environmental protection and technological advancement.

Assume you are designing a city plan. You need parks, but you also need roads. In a similar spirit, regulations are necessary to guarantee that environmental costs be kept to a minimum as our digital world advances. Rules have the authority to do anything, including enact requirements for energy efficiency and the use of renewable energy sources.

One thing a nation may do is enact legislation requiring data centers to disclose their energy usage and carbon emissions. Similar to encouraging drivers to check the emissions of their vehicle, this is a positive step in determining the scope of the issue and developing remedies. Some localities even go so far as to provide tax breaks to data centers that comply with environmental standards, which is similar to providing benefits to owners of electric cars.

However, enforcing rules is a challenging task. Striking a careful balance between restricting innovation and pushing the sector toward less regulated areas while implementing substantial regulatory changes is necessary. Similar to speed limits, it protects people without causing them to be late for work.

International Perspectives on Data Center Energy in Global Discussions

The world offers a patchwork of approaches to controlling data center energy use, all of which take into consideration the distinct requirements and objectives of different countries. Rich in renewable energy, some nations encourage green data centers; energy-poor nations place more emphasis on innovation and efficiency.

Consider Iceland, where data centers can run nearly entirely on renewable energy thanks to geothermal and hydroelectric power. It would be similar to having wind turbines on your lawn and solar panels on your roof providing all of your electricity. Singapore, on the other hand, employs cutting-edge technologies like subsea data centers that use saltwater for cooling, reducing the need for energy-intensive air conditioning, despite its restricted space and lack of renewable sources.

International relationships and agreements are also covered. Similar to how nations are uniting to battle climate change, there is a growing recognition that collaborative efforts to optimize and control data center energy use are crucial. Imagine a situation where a few neighbors band together to pool their resources and equipment in an effort to make their houses more energy-efficient.

This international conversation is essential for addressing the transboundary nature of data flows and the internet, in addition to sharing best practices. Data ultimately crosses national boundaries. Every day, people who are watching films hosted on servers in another country while in another country provide evidence of this interconnection. International cooperation is therefore essential for the digital sector as well as for tackling global issues like climate change.

The Way Forward: Balancing Innovation and Sustainability

Urban planning for a metropolis that is always expanding is similar to the ongoing procedures of data center energy control and management. It calls for vision, adaptability, and the ability to change with the times and with new technology.

The future is probably going to bring more creative energy solutions, more rules, and more international cooperation. The problem is difficult, but the components are there. Sustainable development can be promoted and data center efficiency can be increased via the use of technology and regulatory mechanisms. Integrating these elements presents a challenge: juggling the demands of our data-hungry appetite with the wellbeing of our planet.

Essentially, we have to consider the power and responsibility that come with building the digital infrastructure of the future as we move on with this endeavor. Data centers are about more than simply bits and bandwidth; they're about how we employ these resources to build an offline-online civilization. It is about ensuring that the world we live on does not suffer as our digital progress increases.

CHAPTER SEVEN

Societal Implications

In the digital age, data is becoming just as vital as electricity or water. It controls everything, including our smartphones and banking systems. However, this digital revolution is also accompanied by significant societal issues, particularly in the areas of accessibility, privacy, and power dynamics. The way we address these concerns will determine the destiny of our global community.

The Data Divide's Digital Economy and Accessibility

Envision a society where the key to success lies in knowledge, but not everyone has access to the lock or the key to open it. This is the reality for a large number of people living in the digital age; this phenomenon is frequently referred to as the "data divide." This disparity extends beyond who can afford the newest smartphone; it also involves who can utilize high-speed internet to seek for jobs, who can use online resources to enhance their education, and who is unconnected and hence invisible in the digital economy.

The digital economy is growing quickly, offering never-before-seen opportunities for innovation, entrepreneurship, and job creation. That being said, not everyone profits equally from this economy. Many rural and undeveloped metropolitan areas lack reliable internet connectivity, which makes it difficult for locals to take advantage of healthcare services, start online businesses, or participate in remote learning.

One example of how to break through this barrier in practice is the story of a small hamlet that recently gained access to high-speed internet. Suddenly, local artisans could sell their wares online, doctors could conduct virtual consultations, and students could access educational materials that had not been available before. This change illustrates the important role that digital accessibility plays in fostering community development and individual empowerment.

Power, Data, and Privacy: Handling the New Normal

As our lives become more and more digital, personal data has become the money of the digital era. Companies use the digital traces of every click, like, and share to personalize their products, services, and advertisements. It's not necessarily a bad thing, since this data collection can enhance our online experiences and provide us with content that is catered to our interests. However, it also raises significant questions about power dynamics and privacy.

Imagine going into a store where the salesperson knows your name, your favorite style, and your clothing size without you having to say anything. One way or another, it might streamline and customize your shopping experience. It could seem bothersome, though, to realize that so much about you is known without your consent.

This scenario is consistent with what we've seen online. Thanks to data collection strategies that have created a power imbalance, companies today possess considerably more information about us than we may acknowledge or consent to. The potential for data misuse is a major concern, especially when it comes to sensitive information that could be exploited to change our behavior or gain financial gain.

It's challenging to strike a balance between protecting privacy and leveraging data for creativity in this new normal. Regulations such as the General Data Protection Regulation (GDPR) in the European Union have been good steps in the right direction, aiming to offer consumers more control over their personal data. In order to truly attain privacy in the digital age, it might also be required to alter our cultural views on data ownership and consent.

Building Bridges and Safeguarding Sanctuaries

In the digital age, it's critical to address concerns about privacy and the data divide while still fostering connections and safeguarding safe spaces. Assuring that everyone can benefit from the opportunities provided by the digital economy while respecting each person's right to privacy and control over their personal data is the aim of the strategies being proposed.

One tactic for narrowing the data gap is to use community-driven solutions, such as neighborhood broadband initiatives or public-private partnerships that aim to extend digital infrastructure to underserved areas. These initiatives have the potential to empower local communities by providing them with the necessary resources to participate in the digital economy.

However, safeguarding privacy demands adjustments to both corporate data collection policies and state regulatory frameworks. It involves moving away from an extraction-based data paradigm and toward a stewardship-based one where companies safeguard personal information and prioritize openness and user permission.

Going forward, collaboration will be required from all of us due to the societal implications of the digital age. Governments, businesses, and individuals must work together to ensure that the digital future is inclusive, equitable, and protects our fundamental right to privacy. This is about building a society where advances in technology help people, not the other way around. It goes beyond technology.

Making our way through a continuously changing environment is ultimately what it means to navigate the societal repercussions of data in the digital age. Making choices that ensure everyone has access to opportunities in the digital age, respects people's privacy, and uses data to improve society is crucial. The decisions we make now, as we embark on this journey, will shape the digital pathways of the future and shape the way the digital age remembers us.

CHAPTER EIGHT

Towards a Sustainable Digital Ecosystem

Bringing renewable energy sources into the digital sphere is a critical step in our journey to a sustainable future. This chapter presents the innovative use of artificial intelligence (AI) and machine learning for data center energy efficiency, as well as the success stories of incorporating renewable energy sources. Examining these accomplishments teaches us about the innovations and challenges that stand in the way of a greener future, as well as the areas in which we may improve.

Embracing Renewable Energy: Learnings from Case Studies of Green Change

Reports of successful integration of renewable energy into data centers are starting to appear globally, providing promise for a digital infrastructure that is sustainable. One well-known example is a data center in Sweden that runs entirely on renewable energy, primarily hydroelectric power. This building is an example of how technology and the environment can coexist peacefully since it uses the cold Nordic air to cool its servers in addition to reducing carbon emissions.

Another touching story is about how the building of a solar-powered data center transformed the economy of a tiny community. The initiative created a model for how communities may use solar energy to power digital infrastructure in addition to creating jobs in the area. Rather than being isolated incidents, these success stories reflect a growing trend toward sustainability in the tech sector.

AI and Machine Learning as a Green Catalyst

Data centers now optimize their energy use in a whole different way thanks to the advancements in AI and machine learning. These systems analyze massive amounts of data to forecast and adjust energy consumption, allowing data centers to operate as profitably and effectively as possible.

Let's say a data center's cooling system uses artificial intelligence (AI) to adjust its operation based on server load and the current weather. This is not science fiction; this is reality. These solutions can significantly reduce energy use by just cooling the data center when needed, as opposed to constantly running at full capacity.

Furthermore, operational inefficiencies in the data center that are invisible to humans can be identified by AI algorithms. For example, Google used an AI-based solution to boost the cooling system's efficiency in some of its data centers by 40%. This remarkable achievement demonstrates how artificial intelligence and machine learning may revolutionize energy management in digital infrastructures.

Challenges and Prospective Paths

The road ahead of a fully sustainable digital environment is not without its challenges, notwithstanding the progress made recently. Examples of renewable energy sources that depend on the unpredictable weather are wind and solar power. Because renewable energy sources are unpredictable, advanced energy storage technology and adaptable grid infrastructure must be developed.

Furthermore, the upfront costs of integrating renewable energy sources with AI-based optimization tools may be prohibitive for smaller firms. To get past these financial barriers, the government must provide incentives that encourage investment in green technologies and innovative financing methods.

The Combined Journey

Governments, businesses, and people all need to contribute to the transition to a sustainable digital ecosystem. Enacting legislation that promotes the use of renewable energy sources and the installation of energy-saving devices can be a major contribution from governments. Businesses must prioritize sustainability in their operations since it provides long-term financial benefits as well as environmental advantages.

Additionally, people may influence change by being conscious of their digital footprint and by supporting companies that uphold sustainable business practices. Small actions, including reducing unnecessary data storage and optimizing digital use, can have a significant impact.

To sum up, a future outlook

In light of the intersection of digital innovation and environmental stewardship, the path towards a sustainable digital ecosystem poses both unparalleled possibilities and significant challenges. Incorporating renewable energy sources and optimizing energy consumption strategies for reducing carbon emissions through AI and machine learning are not only beneficial, but they also stand as monuments to human ingenuity and creativity for the benefit of society.

Being sustainable requires a process of innovation, learning, and adjustment. By studying success stories and pushing the boundaries of technology, we can build a digital society that unites us and preserves the earth for coming generations.

By establishing a sustainable digital ecosystem, we are reconsidering the relationship between technology and the environment, going beyond merely solving a technical puzzle. It's a vision of a time where technological innovation and environmental sustainability coexist, allowing us to live digital lives where new technologies are created without compromising the state of the environment. This is not just a fantasy; if we are willing to work together to make it happen, it is a possibility.

Roger Munoz

CHAPTER NINE

Impact of Data Center Boom on Climate Change

The digital revolution is being propelled by the exponential rise of data centers, and the consequences of this boom on the environment are starting to show. Understanding the intricate relationship between climate change and data centers is necessary to navigate the challenges posed by the environmental impact of these facilities. This chapter examines data centers' negative environmental implications, including how they fuel climate change and suggestions for mitigating those effects.

Examining the Environmental Effects: Understanding the Relationship Between Data Centers and Climate Change

Massive amounts of energy are used by data centers, the backbone of our digital infrastructure, to power the servers that handle and store our digital data. Because most of the energy used today comes from fossil fuels, the demand for energy increases greenhouse gas emissions, which worsens climate change. To put that into perspective, it is projected that by 2025, the energy consumed by data centers globally will amount to 1,200 terawatt-hours, or the same amount of electricity consumed by Great Britain as a whole.

Beyond their energy use, data centers have an effect on the environment. The construction and maintenance of data centers requires resources like water and raw materials, which adds to habitat destruction, water pollution, and land degradation. The garbage generated by abandoned servers and electronic equipment contributes to additional environmental problems. A large portion of this waste is disposed of in landfills, where it seeps hazardous substances into the earth and water.

Getting Through the Storm: Methods for Mitigating Data Center-Related Climate Impacts

An integrated approach that considers resource management, energy consumption, and waste management is required to lessen the environmental impact of data centers. One tactic is to power data centers with renewable energy sources. By using solar, wind, hydropower, and other renewable energy sources instead of fossil fuels, data centers can significantly reduce their carbon footprint and reliance on them.

Real-world examples show just how effective this strategy can be. For example, IT giants like Google and Apple have invested heavily in renewable energy projects like solar and wind farms to power their data centers. By sourcing 100% of their energy from renewable sources, these companies reduce their environmental impact and set an example for the industry as a whole.

Another way to lessen the impact of data centers on the climate is to increase energy efficiency. Technologies that can optimize energy utilization and reduce total consumption include virtualization, which allows multiple virtual servers to run on a single physical server, and advanced cooling systems. Data center managers can use these efficiency techniques to lower their operational costs and environmental effect.

Furthermore, lifecycle management strategies can lessen the environmental impact of data center operations. Examples of these strategies include ethical material procurement, recycling, and equipment repair from decommissioned facilities. By extending the life of servers and electrical components and recycling materials, data centers may reduce waste and preserve valuable resources.

Stakeholder cooperation is required to implement these mitigation techniques successfully. Governments, businesses, environmental organizations, and communities as a whole must set ambitious targets for reducing carbon emissions, promoting the use of renewable energy sources, and implementing energy-saving measures in the data center industry.

CHAPTER TEN

Strategies for a Safer Global Environment

In the pursuit of economic and technological advancement, the spread of data centers has become a double-edged sword, offering previously unheard-of possibilities but also posing significant environmental challenges. Achieving a balance between expansion and sustainability is crucial in order to prevent the continued construction of data centers from endangering the health of our planet. In this chapter, we address strategies for achieving this delicate balance and building a more safe global environment.

Understanding the Trade-Offs: Growth vs. Sustainability

The rapid expansion of data centers is a result of the rising demand for digital services such as cloud computing and streaming platforms. Growth promotes innovation and economic progress, but it also severely strains the environment due to rising energy costs and carbon emissions. The challenging element is figuring out how to strike a balance between the need for ongoing expansion and sustainability.

Imagine a crowded city with a rapidly growing population. In addition to boosting the economy and generating jobs, the population expansion strains infrastructure and natural resources. In a similar vein, the expansion of data centers presents opportunities as well as challenges, requiring careful planning and management to minimize its impact on the environment.

Embracing Sustainable Practices: An All-encompassing Approach

To accomplish sustainability in data center development, a complete plan incorporating waste management, resource conservation, energy efficiency, and the use of renewable energy sources is required. By implementing these ideas into every stage of the data center lifecycle, from design and construction to operation and decommissioning, we can minimize the adverse effects on the environment while optimizing the benefits of digital infrastructure.

An important strategy is to design data centers with energy efficiency in mind from the beginning. This means optimizing server topologies, implementing efficient cooling systems, and utilizing energy-saving technology. By reducing their energy use per processing power unit, data centers can simultaneously minimize their operating costs and carbon footprint.

Reaping the Benefits of Renewable Energy

Another key strategy to promote sustainability in data center expansion is to move to renewable energy sources. Solar, wind, hydroelectric, and geothermal power are abundant and clean substitutes for fossil fuels that reduce greenhouse gas emissions and dependency on finite resources. Data centers that engage in renewable energy-powered infrastructure can align their operations with global initiatives aimed at mitigating climate change.

Real-world examples illustrate the strategy's benefits and feasibility. Companies like Microsoft and Facebook have made significant investments in renewable energy, powering their data centers with solar and wind energy. These massive IT companies demonstrate how using renewable energy on a broad scale can have positive environmental effects when data centers grow.

Encouraging Innovation and Collaboration

Innovation and cooperation are two important elements that support sustainability in data center expansion. To develop and implement innovative solutions, we may pool the expertise and assets of governments, businesses, educational institutions, and civil society organizations. This cooperative technique expedites the establishment of a safer society by facilitating policy formation, technological transfer, and knowledge exchange.

Leaders in the industry have teamed up through programs like the carbon Neutral Data Center Pact to attain carbon neutrality in data center operations by 2030.

Signatories to the agreement demonstrate how successful group action can be in addressing environmental challenges by setting ambitious objectives and sharing best practices.

Concluding Discussion: Building a Sustainable Future

In conclusion, despite the rise of data centers, a conscious effort to find a balance between growth and sustainability is required to create a safer global environment. We can create a future that promotes financial success while protecting the environment by adopting sustainable behaviors, using renewable energy sources, and encouraging innovation and teamwork.

Data center development must be handled cautiously and strategically, much like a gardener tends to their garden, preserving its beauty for future generations. By promoting a culture of sustainability and resilience, we can create a future where environmental stewardship and technology innovation coexist, ensuring a safer and more affluent world for everybody.

www.ingramcontent.com/pod-product-compliance
Lightning Source LLC
Chambersburg PA
CBHW070356230526
45471CB00006B/2603